AF266019

This book belongs to:

----

a Warrior Scholar

Kamali Academy

Kamali Academy strives to help warrior parents across the globe provide their young warrior scholars with an education for liberation. This workbook is apart of that continued mission.

Young warrior scholars can use this book and its mental math tips to master basic and advanced division thoroughly and quickly.

Whether you are homeschooling your child or supplementing the education they receive in the public and private school system, this will serve you well.

Parents, be sure to monitor the progress of your warrior scholar and watch them grow.

cover design:
Indigenous Lens

Division

Table of Contents:

Name: ________ Date: __ / __ / __

1 Divide:

$$\begin{array}{r} 9 \\ 2\,\overline{)\,18} \end{array} \quad 2 \times 9 = 18$$

(1) $4\,\overline{)\,36}$ (6) $4\,\overline{)\,24}$ (11) $4\,\overline{)\,8}$ (16) $3\,\overline{)\,27}$

(2) $5\,\overline{)\,40}$ (7) $5\,\overline{)\,15}$ (12) $5\,\overline{)\,5}$ (17) $3\,\overline{)\,21}$

(3) $4\,\overline{)\,20}$ (8) $5\,\overline{)\,25}$ (13) $2\,\overline{)\,2}$ (18) $2\,\overline{)\,10}$

(4) $5\,\overline{)\,10}$ (9) $3\,\overline{)\,18}$ (14) $4\,\overline{)\,24}$ (19) $2\,\overline{)\,16}$

(5) $4\,\overline{)\,4}$ (10) $3\,\overline{)\,27}$ (15) $2\,\overline{)\,8}$ (20) $5\,\overline{)\,40}$

  Kamali Academy Division

Divide: Under each division problem, write the multiplication equation.

(1) $3\overline{)6}$ with quotient 2

$3 \times 2 = 6$

(6) $1\overline{)3}$

(11) $2\overline{)4}$

(16) $3\overline{)21}$

(2) $2\overline{)14}$

(7) $2\overline{)16}$

(12) $1\overline{)7}$

(17) $2\overline{)10}$

(3) $3\overline{)12}$

(8) $1\overline{)8}$

(13) $3\overline{)18}$

(18) $1\overline{)1}$

(4) $1\overline{)6}$

(9) $1\overline{)5}$

(14) $1\overline{)2}$

(19) $4\overline{)12}$

(5) $3\overline{)27}$

(10) $2\overline{)18}$

(15) $1\overline{)4}$

(20) $5\overline{)20}$

2

Name: Date: /   /

1 **Divide:**

(1) $5\overline{)35}$ (6) $2\overline{)18}$ (11) $5\overline{)45}$ (16) $2\overline{)16}$

(2) $4\overline{)16}$ (7) $3\overline{)18}$ (12) $3\overline{)15}$ (17) $2\overline{)12}$

(3) $4\overline{)36}$ (8) $3\overline{)27}$ (13) $5\overline{)30}$ (18) $9\overline{)45}$

(4) $3\overline{)18}$ (9) $3\overline{)12}$ (14) $4\overline{)32}$ (19) $5\overline{)35}$

(5) $5\overline{)25}$ (10) $5\overline{)20}$ (15) $2\overline{)8}$ (20) $2\overline{)18}$

2 **Divide:**

(1) $3\overline{)9}$ (6) $3\overline{)21}$ (11) $3\overline{)24}$ (16) $4\overline{)4}$

(2) $4\overline{)12}$ (7) $2\overline{)10}$ (12) $4\overline{)20}$ (17) $5\overline{)15}$

(3) $5\overline{)25}$ (8) $5\overline{)40}$ (13) $3\overline{)9}$ (18) $5\overline{)10}$

(4) $5\overline{)10}$ (9) $5\overline{)30}$ (14) $4\overline{)12}$ (19) $4\overline{)0}$

(5) $4\overline{)28}$ (10) $4\overline{)24}$ (15) $2\overline{)14}$ (20) $3\overline{)6}$

Name: ___________   Date: ___/___/___

1 Divide:

(1) $2\overline{)8}$ (6) $3\overline{)12}$ (11) $3\overline{)6}$ (16) $3\overline{)3}$

(2) $5\overline{)45}$ (7) $2\overline{)16}$ (12) $6\overline{)18}$ (17) $2\overline{)12}$

(3) $4\overline{)28}$ (8) $2\overline{)14}$ (13) $5\overline{)10}$ (18) $5\overline{)5}$

(4) $8\overline{)24}$ (9) $3\overline{)9}$ (14) $3\overline{)21}$ (19) $3\overline{)27}$

(5) $3\overline{)15}$ (10) $4\overline{)8}$ (15) $2\overline{)18}$ (20) $3\overline{)12}$

     Kamali Academy Division

(1) 2)$\overline{1\,4}$

(2) 5)$\overline{2\,0}$

(3) 4)$\overline{3\,2}$

(4) 4)$\overline{8}$

(5) 4)$\overline{2\,4}$

(6) 3)$\overline{3}$

(7) 2)$\overline{4}$

(8) 5)$\overline{3\,0}$

(9) 3)$\overline{1\,5}$

(10) 4)$\overline{1\,6}$

(11) 3)$\overline{1\,8}$

(12) 4)$\overline{4}$

(13) 4)$\overline{3\,6}$

(14) 2)$\overline{1\,2}$

(15) 4)$\overline{1\,2}$

(16) 4)$\overline{2\,8}$

(17) 5)$\overline{4\,0}$

(18) 3)$\overline{9}$

(19) 2)$\overline{1\,0}$

(20) 5)$\overline{5}$

Name: Date: /   /

1 Divide:

(1) $4\overline{)16}$ (6) $5\overline{)20}$ (11) $2\overline{)6}$ (16) $4\overline{)24}$

(2) $2\overline{)10}$ (7) $2\overline{)4}$ (12) $5\overline{)15}$ (17) $3\overline{)12}$

(3) $3\overline{)9}$ (8) $4\overline{)12}$ (13) $2\overline{)16}$ (18) $4\overline{)20}$

(4) $2\overline{)14}$ (9) $2\overline{)2}$ (14) $2\overline{)8}$ (19) $3\overline{)3}$

(5) $8\overline{)16}$ (10) $2\overline{)12}$ (15) $2\overline{)18}$ (20) $6\overline{)24}$

     Kamali Academy Division

(1) $4\overline{)20}$ (6) $2\overline{)6}$ (11) $5\overline{)30}$ (16) $2\overline{)14}$

(2) $5\overline{)40}$ (7) $4\overline{)4}$ (12) $3\overline{)9}$ (17) $5\overline{)15}$

(3) $5\overline{)45}$ (8) $5\overline{)20}$ (13) $5\overline{)5}$ (18) $3\overline{)15}$

(4) $4\overline{)16}$ (9) $3\overline{)21}$ (14) $5\overline{)25}$ (19) $2\overline{)16}$

(5) $5\overline{)35}$ (10) $4\overline{)24}$ (15) $3\overline{)18}$ (20) $5\overline{)10}$

Read the problem carefully and solve. Show your work under each question.

---

1. The bookcase has 24 books in it. There are 2 shelves in the bookcase. How many books are in each row?

   The bookcase has _________________ books.

   There are ____________ shelves in the bookcase.

   There are ____________ books on each shelf.

2. Noliwe has 15 sticks of bubble gum and 12 sticks of peppermint gum. If she gives each of her 3 friends the same number of sticks of gum, how many sticks of gum will each of Noliwe's friends have?

   Each of Noliwe's friends will have ____________ sticks of gum.

3. Akua earned 27 dollars for making 3 pairs of earrings on Saturday. Akua earned the same amount of money for each pair of earrings. How much did she earn for each pair?

   Akua earned ____________ dollars for each pair of earrings she made.

4. Kwame brought 30 Black Panther cupcakes to share with his class for his birthday. He arranged the cupcakes by flavor into 5 different groups with the same amount of cupcakes in each group. How many cupcakes are in each group?

   There are ____________ cupcakes in each group.

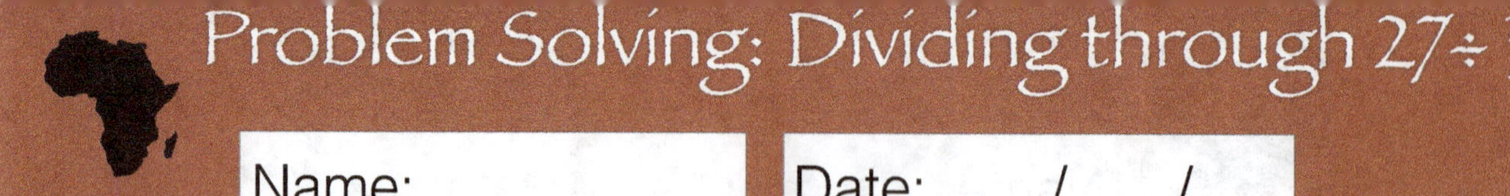

Name: _______________   Date: ___ / ___ / ___

Read the problem carefully and solve. Show your work under each question.

Jomo is opening a pet store. He has 9 turtles, 14 snails, and 24 goldfish that he will put into fish tanks around the store.

---

1. Jomo has 3 tanks for turtles. He wants to put the same number in each tank. How many turtles will he put in each tank?

_______________________ turtles

2. Jomo has 2 tanks for snails. He wants to put the same number of snails in each tank. How many snails will he put in each tank?

_______________________ snails

3. Jomo has 3 tanks for goldfish. If Jomo puts the same number of goldfish in each tank, how many goldfish will be in each tank?

_______________________ goldfish

Name: Date: /   /

Read the problem carefully and solve. Show your work under each question.

Ida helps get 4 sailboats ready for a sailing class. She divides the supplies evenly between each sailboat. She has 8 sails and 20 life jackets. All of the sailboats need new ropes for their sails. There are 16 pieces of rope.

1. Ida puts the same number of sails on each boat. How many sails does she put on each sailboat?

_______________ sails

2. The pieces of rope are evenly divided among the boats. How many pieces of rope does each sailboat get?

_______________ pieces of rope

3. Ida puts the same number of life jackets on each boat. How many life jackets does she put on each boat?

_______________ life jackets

4. The next sailing class plans to use 8 boats. Ida learns that she will need 32 life jackets. How many jackets will each boat get?

_______________ life jackets

Division Practice

Name: _______________   Date: /   /

Read the problem carefully and solve. Show your work under each question.

---

1. Jomo's fish store has 18 goldfish. The fish are in 3 aquariums. The same number of goldfish are in each aquarium. How many goldfish are in each aquarium?

There are __________ goldfish.

There are __________ aquariums.

There are __________ goldfish in each aquarium.

2. Akili has 16 shoes in her closet. A pair of shoes is a group of 2 shoes. How many pairs of shoes does Akili have?

Akili has __________ shoes.

A pair is a group of __________ shoes.

Akili has __________ pairs of shoes.

Division Practice

Name: ___________   Date: ___ / ___ / ___

Read the problem carefully and solve. Show your work under each question.

---

1. Eight people paid a total of 24 dollars for admission into the school festival. If each ticket cost the same amount, how much did each ticket cost?

   The cost of each ticket was ___________ dollars.

2. A family of 5 takes an ice chest to the beach. There are 10 water bottles in the ice chest. How many water bottles will each person receive if each person receives the same number of water bottles?

   Each person will receive ___________ water bottles.

3. Eighteen potatoes were harvested in the garden. If each person in the garden harvested 2 potatoes, how many were in the garden? Draw your answer.

   There were ___________ people in the garden.

   XX   XX   XX   XX   XX   XX   XX   XX   XX

4. A farmer wants to share 28 tomatoes and 8 cucumbers with 4 neighbors. How many total vegetables will each neighbor receive?

   Each neighbor receives ___________ vegetables.

Name: ___________   Date: /   /

1 Divide:

(1) 4 ⟌ 36

(2) 6 ⟌ 54

(3) 4 ⟌ 8

(4) 8 ⟌ 16

(5) 2 ⟌ 12

(6) 10 ⟌ 30

(7) 9 ⟌ 81

(8) 4 ⟌ 4

(9) 6 ⟌ 30

(10) 3 ⟌ 9

(11) 7 ⟌ 14

(12) 3 ⟌ 21

(13) 5 ⟌ 40

(14) 3 ⟌ 24

(15) 4 ⟌ 16

(16) 5 ⟌ 20

(17) 10 ⟌ 100

(18) 7 ⟌ 7

(19) 3 ⟌ 27

(20) 5 ⟌ 35

(1) 6 | 3 6   (6) 8 | 7 2   (11) 7 | 2 8   (16) 5 | 3 0

(2) 4 | 2 8   (7) 6 | 4 8   (12) 5 | 4 0   (17) 2 | 1 8

(3) 5 | 3 5   (8) 7 | 6 3   (13) 6 | 6   (18) 4 | 1 2

(4) 6 | 2 4   (9) 5 | 4 5   (14) 7 | 5 6   (19) 5 | 2 5

(5) 3 | 2 1   (10) 6 | 1 2   (15) 5 | 4 5   (20) 7 | 2 1

Name: _____________ Date: ___ / ___ / ___

1 Divide:

(1) $4 \overline{)28}$ (6) $4 \overline{)20}$ (11) $6 \overline{)30}$ (16) $8 \overline{)56}$

(2) $2 \overline{)16}$ (7) $7 \overline{)35}$ (12) $7 \overline{)56}$ (17) $6 \overline{)30}$

(3) $6 \overline{)54}$ (8) $5 \overline{)30}$ (13) $2 \overline{)14}$ (18) $8 \overline{)64}$

(4) $7 \overline{)63}$ (9) $4 \overline{)12}$ (14) $6 \overline{)36}$ (19) $7 \overline{)35}$

(5) $6 \overline{)24}$ (10) $4 \overline{)16}$ (15) $4 \overline{)32}$ (20) $6 \overline{)36}$

          Kamali Academy Division

Divide:

(1) 9 ⟌ 3 6   (6) 9 ⟌ 1 8   (11) 7 ⟌ 4 9   (16) 9 ⟌ 7 2

(2) 9 ⟌ 8 1   (7) 7 ⟌ 1 4   (12) 8 ⟌ 1 6   (17) 8 ⟌ 4 0

(3) 7 ⟌ 5 6   (8) 7 ⟌ 2 1   (13) 9 ⟌ 2 7   (18) 8 ⟌ 2 4

(4) 5 ⟌ 3 5   (9) 6 ⟌ 4 8   (14) 9 ⟌ 9   (19) 6 ⟌ 4 8

(5) 8 ⟌ 2 4   (10) 9 ⟌ 4 5   (15) 5 ⟌ 4 2   (20) 9 ⟌ 5 4

Name: Date: /   /

1st: Find out how many numbers will be in the answer by multiplying by 10, 100, 1000, & so on.

$2 \times 10 = 20$
$2 \times \underline{100} = 200$
$2 \times \underline{1000} = 2000$

Since $2 \times 1000 = 2000$ and 2000 is greater than 246, the answer will be between 100 and 999, a 3-digit number.

$$2\,\overline{)246}$$

x	100	200	300
2	200	400	1200

246 is between 200 and 400. $246 \div 2$ is between 100 and 200. The hundreds digit is 1.

$$
\begin{array}{r}
1\,x\,x \\
2\,\overline{)246} \\
-200 \\
\hline
46
\end{array}
$$
$2 \times 100 = 200$

subtract

x	10	20	30	40
2	20	40	60	80

46 is between 40 and 60. $46 \div 2$ is between 20 and 40. The tens digit is 2.

$$
\begin{array}{r}
1\,2\,x \\
2\,\overline{)246} \\
-200 \\
\hline
46 \\
-40 \\
\hline
6
\end{array}
$$
$2 \times 20 = 40$

subtract

x	1	2	3	4
2	2	4	6	8

$6 \div 2$ is 3. The ones digit is 3.

$$
\begin{array}{r}
1\,2\,3 \\
2\,\overline{)246} \\
-200 \\
\hline
46 \\
-40 \\
\hline
6 \\
-6 \\
\hline
0
\end{array}
$$

The answer is 123.

There is no remainder.

$2 \times 3 = 6$

subtract

 **Try it!**

$$3\,\overline{)636}$$

1st: Find out how many numbers will be in the answer by multiplying by 10, 100, 1000, & so on.

8 x <u>10</u> = 80
8 x <u>100</u> = 800

Since 8 x 100 = 800 and 800 is greater than 453, the answer will be between 10 and 99, a 2-digit number.

8 | 453

x	10	20	30	40	50	60
8	80	160	240	320	400	480

453 is between 400 and 480. 453 ÷ 8 is between 50 and 60. The tens digit is 5.

```
        5 x
   8 | 4 5 3
      -4 0 0        8 x 50 = 400
      ───────
        5 3         subtract
```

x	1	2	3	4	5	6	7
8	8	16	24	32	40	48	56

53 is between 48 and 56. 53 ÷ 8 is between 6 and 7. The ones digit is 6.

```
        5 6 r 5
   8 | 4 5 3
      -4 0 0
      ───────
        5 3
       - 4 8        8 x 6 = 48
      ───────
          5         subtract
```

 **Try it!**

4 | 1 7 3

 **Try it!**

7 | 5 1 1

Name: Date: /   /

1 Divide:

(1) $2\,\overline{)12}$ (6) $7\,\overline{)28}$ (11) $4\,\overline{)26}$ (16) $8\,\overline{)46}$

(2) $3\,\overline{)18}$ (7) $8\,\overline{)32}$ (12) $5\,\overline{)26}$ (17) $8\,\overline{)70}$

(3) $4\,\overline{)24}$ (8) $9\,\overline{)36}$ (13) $5\,\overline{)33}$ (18) $9\,\overline{)76}$

(4) $5\,\overline{)25}$ (9) $2\,\overline{)15}$ (14) $6\,\overline{)37}$ (19) $3\,\overline{)99}$

(5) $6\,\overline{)30}$ (10) $3\,\overline{)20}$ (15) $7\,\overline{)44}$ (20) $3\,\overline{)87}$

2 **Divide:**

(1) 2 ⟌ 2 6   (6) 2 ⟌ 4 6   (11) 2 ⟌ 3 0   (16) 3 ⟌ 3 9

(2) 2 ⟌ 2 8   (7) 2 ⟌ 4 8   (12) 2 ⟌ 3 2   (17) 3 ⟌ 6 0

(3) 2 ⟌ 4 0   (8) 2 ⟌ 6 0   (13) 2 ⟌ 3 6   (18) 3 ⟌ 6 3

(4) 2 ⟌ 4 2   (9) 2 ⟌ 6 4   (14) 2 ⟌ 5 2   (19) 3 ⟌ 6 9

(5) 2 ⟌ 4 4   (10) 2 ⟌ 8 0   (15) 3 ⟌ 3 6   (20) 3 ⟌ 4 2

Name: Date: /   /

1   Divide:

(1)   3 ⟌ 2 4   (6)   3 ⟌ 7 5   (11)   4 ⟌ 7 6   (16)   4 ⟌ 5 6

(2)   3 ⟌ 3 3   (7)   4 ⟌ 4 4   (12)   4 ⟌ 8 0   (17)   4 ⟌ 8 4

(3)   3 ⟌ 4 8   (8)   4 ⟌ 4 8   (13)   4 ⟌ 7 2   (18)   4 ⟌ 9 6

(4)   3 ⟌ 5 1   (9)   4 ⟌ 5 2   (14)   4 ⟌ 6 0   (19)   4 ⟌ 3 2

(5)   3 ⟌ 6 9   (10)   4 ⟌ 6 4   (15)   4 ⟌ 6 8   (20)   4 ⟌ 2 4

② Divide:

(1) $5\overline{)15}$	(6) $5\overline{)65}$
(2) $5\overline{)25}$	(7) $5\overline{)80}$
(3) $5\overline{)40}$	(8) $5\overline{)90}$
(4) $5\overline{)75}$	(9) $5\overline{)85}$
(5) $5\overline{)55}$	(10) $5\overline{)95}$

(11) $6\overline{)24}$	(16) $6\overline{)84}$
(12) $6\overline{)30}$	(17) $6\overline{)78}$
(13) $6\overline{)48}$	(18) $6\overline{)96}$
(14) $6\overline{)66}$	(19) $6\overline{)90}$
(15) $6\overline{)72}$	(20) $3\overline{)51}$

Name:

Date: / /

1 Divide:

(1) 2) 5 6   (6) 3) 6 0   (11) 7) 6 3   (16) 2) 7 0

(2) 4) 5 6   (7) 4) 6 0   (12) 9) 6 3   (17) 5) 7 0

(3) 7) 5 6   (8) 5) 6 0   (13) 2) 6 6   (18) 7) 7 0

(4) 8) 5 6   (9) 6) 6 0   (14) 3) 6 6   (19) 2) 7 2

(5) 2) 6 0   (10) 3) 6 3   (15) 6) 6 6   (20) 4) 7 2

(1) 2⟌78   (6) 5⟌80   (11) 4⟌84   (16) 5⟌90

(2) 3⟌78   (7) 8⟌80   (12) 6⟌84   (17) 6⟌90

(3) 6⟌78   (8) 3⟌81   (13) 7⟌84   (18) 9⟌90

(4) 2⟌80   (9) 9⟌81   (14) 2⟌90   (19) 2⟌96

(5) 4⟌80   (10) 3⟌84   (15) 3⟌90   (20) 3⟌96

Name: ___________ Date: ___ / ___ / ___

1 **Divide:**

(1) 2 ⟌ 1 3

(2) 2 ⟌ 2 5

(3) 2 ⟌ 3 1

(4) 2 ⟌ 4 3

(5) 2 ⟌ 5 7

(6) 2 ⟌ 6 1

(7) 3 ⟌ 2 6

(8) 3 ⟌ 3 7

(9) 3 ⟌ 4 4

(10) 3 ⟌ 5 3

(11) 3 ⟌ 5 5

(12) 3 ⟌ 6 1

(13) 3 ⟌ 6 5

(14) 4 ⟌ 3 2

(15) 4 ⟌ 3 8

(16) 4 ⟌ 4 7

(17) 4 ⟌ 5 3

(18) 4 ⟌ 6 0

(19) 4 ⟌ 6 1

(20) 4 ⟌ 6 5

② Divide:

(1) $2\overline{)55}$	(6) $3\overline{)62}$	(11) $5\overline{)69}$	(16) $7\overline{)74}$
(2) $2\overline{)63}$	(7) $4\overline{)49}$	(12) $5\overline{)71}$	(17) $8\overline{)74}$
(3) $3\overline{)34}$	(8) $4\overline{)60}$	(13) $5\overline{)73}$	(18) $9\overline{)74}$
(4) $3\overline{)47}$	(9) $4\overline{)66}$	(14) $5\overline{)74}$	(19) $6\overline{)75}$
(5) $3\overline{)58}$	(10) $5\overline{)67}$	(15) $6\overline{)74}$	(20) $7\overline{)75}$

Name: **Date:**   /   /

1   Divide:

(1) 2) 7 6	(6) 8) 7 6	(11) 5) 7 7	(16) 2) 7 8
(2) 4) 7 6	(7) 9) 7 6	(12) 6) 7 7	(17) 3) 7 8
(3) 5) 7 6	(8) 2) 7 7	(13) 7) 7 7	(18) 4) 7 8
(4) 6) 7 6	(9) 3) 7 7	(14) 8) 7 7	(19) 5) 7 8
(5) 7) 7 6	(10) 4) 7 7	(15) 9) 7 7	(20) 7) 7 8

Divide:

(1) $6\overline{)79}$ (6) $3\overline{)80}$ (11) $8\overline{)80}$ (16) $5\overline{)82}$

(2) $7\overline{)79}$ (7) $4\overline{)80}$ (12) $9\overline{)80}$ (17) $6\overline{)82}$

(3) $8\overline{)79}$ (8) $5\overline{)80}$ (13) $2\overline{)82}$ (18) $7\overline{)82}$

(4) $9\overline{)79}$ (9) $6\overline{)80}$ (14) $3\overline{)82}$ (19) $8\overline{)82}$

(5) $2\overline{)80}$ (10) $7\overline{)80}$ (15) $4\overline{)82}$ (20) $9\overline{)82}$

Name:

Date: / /

1 Divide:

(1) $2\overline{)85}$ (6) $8\overline{)85}$ (11) $5\overline{)87}$ (16) $2\overline{)89}$

(2) $4\overline{)85}$ (7) $9\overline{)85}$ (12) $6\overline{)87}$ (17) $3\overline{)89}$

(3) $5\overline{)85}$ (8) $2\overline{)87}$ (13) $7\overline{)87}$ (18) $4\overline{)89}$

(4) $6\overline{)85}$ (9) $3\overline{)87}$ (14) $8\overline{)87}$ (19) $5\overline{)89}$

(5) $7\overline{)85}$ (10) $4\overline{)87}$ (15) $9\overline{)87}$ (20) $7\overline{)89}$

Divide:

(1) $6\overline{)90}$	(6) $3\overline{)91}$	(11) $8\overline{)91}$	(16) $5\overline{)93}$
(2) $7\overline{)90}$	(7) $4\overline{)91}$	(12) $9\overline{)91}$	(17) $6\overline{)93}$
(3) $8\overline{)90}$	(8) $5\overline{)91}$	(13) $2\overline{)93}$	(18) $7\overline{)93}$
(4) $9\overline{)90}$	(9) $6\overline{)91}$	(14) $3\overline{)93}$	(19) $8\overline{)93}$
(5) $2\overline{)91}$	(10) $7\overline{)91}$	(15) $4\overline{)93}$	(20) $9\overline{)94}$

Name: Date: /   /

1st: Find out how many numbers will be in the answer by multiplying by 10, 100, 1000, & so on.

$8 \times \underline{10} = 80$

$8 \times \underline{100} = 800$

Since $8 \times 100 = 800$ and 800 is greater than 453, the answer will be between 10 and 99, a 2-digit number.

$$8 \overline{)453}$$

x	10	20	30	40	50	60
8	80	160	240	320	400	480

453 is between 400 and 480. $453 \div 8$ is between 50 and 60. The tens digit is 5.

$$\begin{array}{r} 5\,x \\ 8\,\overline{)453} \\ -400 \\ \hline 53 \end{array} \quad \begin{array}{l} 8 \times 50 = 400 \\ \\ \text{subtract} \end{array}$$

x	1	2	3	4	5	6	7
8	8	16	24	32	40	48	56

53 is between 48 and 56. $53 \div 8$ is between 6 and 7. The ones digit is 6.

$$\begin{array}{r} 56 \text{ r } 5 \\ 8\,\overline{)453} \\ -400 \\ \hline 53 \\ -48 \\ \hline 5 \end{array} \quad \begin{array}{l} \\ \\ \\ 8 \times 6 = 48 \\ \\ \text{subtract} \end{array}$$

 **Try it!**

$$4 \overline{)173}$$

 **Try it!**

$$7 \overline{)511}$$

      Kamali Academy Division

Name: Date: /   /

1 **Divide:**

(1) $2\overline{)246}$ (6) $3\overline{)369}$ (11) $2\overline{)228}$ (16) $3\overline{)345}$

(2) $2\overline{)240}$ (7) $3\overline{)639}$ (12) $2\overline{)230}$ (17) $3\overline{)657}$

(3) $2\overline{)624}$ (8) $3\overline{)960}$ (13) $2\overline{)232}$ (18) $3\overline{)675}$

(4) $2\overline{)846}$ (9) $4\overline{)448}$ (14) $2\overline{)458}$ (19) $4\overline{)456}$

(5) $2\overline{)840}$ (10) $4\overline{)840}$ (15) $2\overline{)636}$ (20) $4\overline{)472}$

(1) $2 \overline{)120}$ (6) $8 \overline{)120}$ (11) $5 \overline{)130}$ (16) $4 \overline{)156}$

(2) $3 \overline{)120}$ (7) $9 \overline{)126}$ (12) $7 \overline{)140}$ (17) $6 \overline{)168}$

(3) $4 \overline{)120}$ (8) $2 \overline{)128}$ (13) $8 \overline{)144}$ (18) $7 \overline{)168}$

(4) $5 \overline{)120}$ (9) $3 \overline{)126}$ (14) $9 \overline{)144}$ (19) $8 \overline{)168}$

(5) $6 \overline{)120}$ (10) $4 \overline{)128}$ (15) $3 \overline{)156}$ (20) $9 \overline{)171}$

Name: Date: /   /

1 **Divide:**

(1) $2\overline{)224}$ (6) $9\overline{)261}$ (11) $9\overline{)378}$ (16) $8\overline{)464}$

(2) $4\overline{)228}$ (7) $3\overline{)348}$ (12) $2\overline{)436}$ (17) $9\overline{)468}$

(3) $6\overline{)234}$ (8) $5\overline{)345}$ (13) $4\overline{)460}$ (18) $2\overline{)512}$

(4) $7\overline{)238}$ (9) $7\overline{)364}$ (14) $6\overline{)462}$ (19) $3\overline{)513}$

(5) $8\overline{)256}$ (10) $8\overline{)368}$ (15) $7\overline{)462}$ (20) $4\overline{)524}$

     Kamali Academy Division

(1) $7\overline{)616}$ (6) $5\overline{)630}$ (11) $3\overline{)318}$ (16) $4\overline{)812}$

(2) $8\overline{)608}$ (7) $7\overline{)630}$ (12) $3\overline{)618}$ (17) $5\overline{)515}$

(3) $9\overline{)612}$ (8) $9\overline{)630}$ (13) $3\overline{)624}$ (18) $6\overline{)618}$

(4) $2\overline{)630}$ (9) $3\overline{)300}$ (14) $4\overline{)412}$ (19) $7\overline{)721}$

(5) $4\overline{)660}$ (10) $3\overline{)306}$ (15) $2\overline{)618}$ (20) $8\overline{)816}$

Name: Date: /   /

1 Divide:

(1)   2) 612   (6)   9) 756   (11)   9) 810   (16)   8) 720

(2)   4) 756   (7)   3) 816   (12)   2) 140   (17)   9) 729

(3)   6) 612   (8)   5) 820   (13)   4) 428   (18)   2) 818

(4)   7) 756   (9)   7) 819   (14)   6) 636   (19)   3) 903

(5)   8) 752   (10)   8) 824   (15)   7) 686   (20)   5) 945

     Kamali Academy Division

2 **Divide:**

(1) $6\overline{)852}$	(6) $4\overline{)576}$	(11) $5\overline{)525}$	(16) $5\overline{)380}$
(2) $7\overline{)763}$	(7) $7\overline{)742}$	(12) $6\overline{)330}$	(17) $6\overline{)240}$
(3) $9\overline{)945}$	(8) $8\overline{)832}$	(13) $8\overline{)416}$	(18) $7\overline{)735}$
(4) $2\overline{)416}$	(9) $3\overline{)438}$	(14) $9\overline{)234}$	(19) $8\overline{)848}$
(5) $3\overline{)321}$	(10) $4\overline{)636}$	(15) $4\overline{)392}$	(20) $9\overline{)963}$

Name: ___________ Date: ___ / ___ / ___

1 **Divide:**

(1) 2) 6 1 6   (6) 9) 6 7 5   (11) 9) 6 9 3   (16) 8) 7 1 2

(2) 4) 6 3 6   (7) 3) 6 7 8   (12) 2) 6 9 6   (17) 9) 7 2 0

(3) 6) 6 4 2   (8) 5) 6 8 0   (13) 4) 6 9 6   (18) 2) 7 2 0

(4) 7) 6 4 4   (9) 7) 6 8 4   (14) 6) 7 0 2   (19) 3) 7 2 0

(5) 8) 6 7 2   (10) 8) 6 8 8   (15) 7) 7 1 4   (20) 5) 7 2 0

Kamali Academy Division

(1) $6\overline{)720}$ (6) $4\overline{)752}$ (11) $4\overline{)788}$ (16) $3\overline{)804}$

(2) $7\overline{)728}$ (7) $7\overline{)760}$ (12) $5\overline{)790}$ (17) $4\overline{)812}$

(3) $9\overline{)738}$ (8) $8\overline{)768}$ (13) $6\overline{)792}$ (18) $7\overline{)812}$

(4) $2\overline{)748}$ (9) $3\overline{)774}$ (14) $8\overline{)800}$ (19) $8\overline{)816}$

(5) $3\overline{)750}$ (10) $4\overline{)786}$ (15) $9\overline{)801}$ (20) $9\overline{)819}$

Name: Date: /   /

1 **Divide:**

(1) $2\overline{)820}$ (6) $2\overline{)846}$ (11) $9\overline{)864}$ (16) $9\overline{)873}$

(2) $4\overline{)824}$ (7) $3\overline{)846}$ (12) $3\overline{)864}$ (17) $2\overline{)874}$

(3) $6\overline{)828}$ (8) $5\overline{)850}$ (13) $6\overline{)870}$ (18) $3\overline{)876}$

(4) $7\overline{)833}$ (9) $6\overline{)852}$ (14) $7\overline{)875}$ (19) $4\overline{)876}$

(5) $8\overline{)840}$ (10) $7\overline{)854}$ (15) $8\overline{)872}$ (20) $5\overline{)880}$

Divide:

(1) $2\overline{)871}$ (6) $6\overline{)333}$ (11) $6\overline{)401}$ (16) $3\overline{)517}$

(2) $2\overline{)457}$ (7) $7\overline{)333}$ (12) $7\overline{)401}$ (17) $5\overline{)517}$

(3) $3\overline{)458}$ (8) $8\overline{)333}$ (13) $8\overline{)401}$ (18) $6\overline{)517}$

(4) $4\overline{)634}$ (9) $3\overline{)401}$ (14) $9\overline{)401}$ (19) $7\overline{)517}$

(5) $5\overline{)333}$ (10) $4\overline{)401}$ (15) $2\overline{)517}$ (20) $9\overline{)517}$

Name: Date: /   /

1 **Divide:**

(1) $2\,\overline{)576}$ (6) $9\,\overline{)576}$ (11) $9\,\overline{)601}$ (16) $8\,\overline{)777}$

(2) $4\,\overline{)576}$ (7) $3\,\overline{)601}$ (12) $2\,\overline{)777}$ (17) $9\,\overline{)777}$

(3) $6\,\overline{)576}$ (8) $5\,\overline{)601}$ (13) $4\,\overline{)777}$ (18) $2\,\overline{)801}$

(4) $7\,\overline{)576}$ (9) $7\,\overline{)601}$ (14) $6\,\overline{)777}$ (19) $3\,\overline{)801}$

(5) $8\,\overline{)576}$ (10) $8\,\overline{)601}$ (15) $7\,\overline{)777}$ (20) $5\,\overline{)801}$

(1) $7\overline{)801}$ (6) $7\overline{)853}$ (11) $4\overline{)581}$ (16) $4\overline{)422}$

(2) $8\overline{)801}$ (7) $8\overline{)853}$ (12) $6\overline{)613}$ (17) $5\overline{)519}$

(3) $9\overline{)801}$ (8) $9\overline{)853}$ (13) $8\overline{)468}$ (18) $6\overline{)256}$

(4) $3\overline{)853}$ (9) $2\overline{)441}$ (14) $9\overline{)367}$ (19) $7\overline{)732}$

(5) $5\overline{)853}$ (10) $3\overline{)628}$ (15) $3\overline{)320}$ (20) $9\overline{)260}$

4-Digits ÷ 1-Digit

Name: Date: /   /

1   **Divide:**

(1) $2\overline{)1246}$ (5) $9\overline{)2052}$ (9) $9\overline{)3633}$

(2) $5\overline{)1205}$ (6) $3\overline{)2529}$ (10) $2\overline{)4160}$

(3) $6\overline{)1200}$ (7) $6\overline{)2562}$ (11) $4\overline{)4076}$

(4) $7\overline{)1673}$ (8) $4\overline{)3632}$ (12) $6\overline{)4144}$

      Kamali Academy Division

2 **Divide:**

(1)   2 ⟌ 5 0 3 0

(2)   4 ⟌ 5 0 3 6

(3)   6 ⟌ 5 0 3 4

(4)   3 ⟌ 7 0 3 2

(5)   5 ⟌ 7 0 3 5

(6)   7 ⟌ 7 0 3 5

(7)   9 ⟌ 8 0 0 1

(8)   2 ⟌ 8 0 1 2

(9)   6 ⟌ 8 0 0 4

(10)   7 ⟌ 8 0 0 1

(11)   3 ⟌ 9 0 0 6

(12)   9 ⟌ 9 0 6 8

Name: Date: /   /

1 **Divide:**

(1) $8\overline{)5224}$

(2) $6\overline{)4608}$

(3) $4\overline{)2686}$

(4) $2\overline{)1091}$

(5) $7\overline{)3602}$

(6) $3\overline{)1747}$

(7) $4\overline{)5817}$

(8) $6\overline{)6133}$

(9) $8\overline{)5000}$

(10) $9\overline{)3679}$

(11) $3\overline{)3103}$

(12) $5\overline{)5193}$

     Kamali Academy Division

(1) 8 | 2 8 7 2

(2) 6 | 1 2 1 8

(3) 4 | 3 6 4 0

(4) 9 | 6 1 3 8

(5) 7 | 3 2 7 6

(6) 5 | 2 3 4 0

(7) 8 | 7 5 7 6

(8) 4 | 6 1 3 2

(9) 3 | 6 0 7 8

(10) 8 | 6 9 3 5

(11) 7 | 5 9 8 9

(12) 4 | 3 4 5 8

Name: Date: /   /

1st: Find out how many numbers will be in the answer by multiplying by 10, 100, 1000, & so on.

22 x <u>10</u> = 220

Since 22 x 10 = 220 and 220 is greater than 145, the answer will be between 1 and 9, a 1-digit number.

22 | 1 4 5

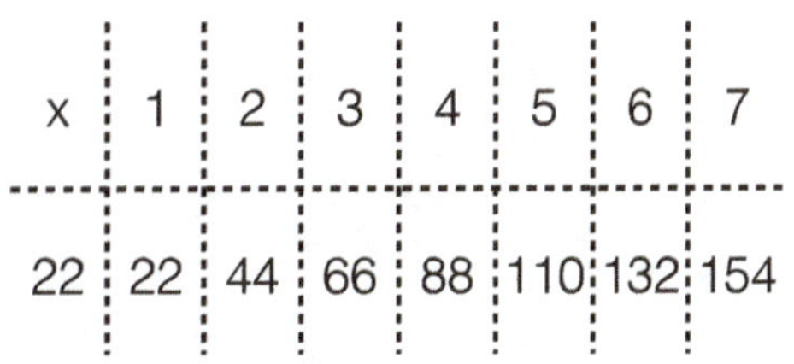

x	1	2	3	4	5	6	7
22	22	44	66	88	110	132	154

145 is between 132 and 154. 145 ÷ 22 is between 6 and 7. The ones digit is 6.

```
        6  r 13
22 | 1 4 5
    -1 3 2        22 x 6 = 132
    -------
      1 3      subtract
```

1st: Find out how many numbers will be in the answer by multiplying by 10, 100, 1000, & so on.

42 x <u>10</u> = 420
42 x 100 = 4200

Since 42 x 100 = 4200 and 4200 is greater than 672, the answer will be between 10 and 99, a 2-digit number.

42 | 6 7 2

x	10	20	30	40
42	420	840	1260	1680

672 is between 420 and 840. 672 ÷ 42 is between 10 and 20. The tens digit is 1.

```
         1 x
42 | 6 7 2
    -4 2 0        42 x 10 = 420
    -------
      2 5 2     subtract
```

x	1	2	3	4	5	6	7
42	42	84	126	168	210	252	294

252 ÷ 42 is 6. The ones digit is 6.

```
        1 6
42 | 6 7 2
    -4 2 0
    -------
      2 5 2
     -2 5 2        42 x 6 = 252
    -------
          0     subtract
```

Name: ___________   Date: ___ / ___ / ___

1 **Divide:**

(1) 21 | 4 5

(2) 21 | 4 7

(3) 21 | 4 8

(4) 21 | 4 9

(5) 21 | 6 5

(6) 21 | 6 9

(7) 21 | 7 0

(8) 21 | 7 2

(9) 21 | 7 8

(10) 21 | 8 5

2 **Divide:**

(1) $21\overline{)86}$

(2) $21\overline{)87}$

(3) $21\overline{)89}$

(4) $21\overline{)94}$

(5) $21\overline{)99}$

(6) $21\overline{)107}$

(7) $21\overline{)109}$

(8) $21\overline{)127}$

(9) $21\overline{)129}$

(10) $21\overline{)119}$

Name: Date: /   /

1   Divide:

(1)   21$\overline{)88}$

(2)   21$\overline{)84}$

(3)   21$\overline{)83}$

(4)   21$\overline{)90}$

(5)   21$\overline{)82}$

(6)   21$\overline{)95}$

(7)   21$\overline{)108}$

(8)   21$\overline{)105}$

(9)   21$\overline{)102}$

(10)   21$\overline{)110}$

(1) 21 | 1 3 0

(2) 21 | 1 3 5

(3) 21 | 1 4 8

(4) 21 | 1 4 7

(5) 21 | 1 4 5

(6) 21 | 1 6 8

(7) 21 | 1 6 5

(8) 21 | 1 7 8

(9) 21 | 1 8 9

(10) 21 | 2 0 4

Name: Date: /   /

1 **Divide:**

(1) 31 | 6 5

(2) 31 | 7 5

(3) 31 | 9 9

(4) 31 | 9 0

(5) 31 | 9 5

(6) 31 | 1 1 0

(7) 31 | 1 2 5

(8) 31 | 1 2 0

(9) 31 | 1 3 0

(10) 31 | 1 4 0

  Kamali Academy Division

Divide:

(1) $31 \overline{)145}$ (5) $31 \overline{)190}$ (9) $31 \overline{)250}$

(2) $31 \overline{)155}$ (6) $31 \overline{)215}$ (10) $31 \overline{)300}$

(3) $31 \overline{)150}$ (7) $31 \overline{)220}$

(4) $31 \overline{)185}$ (8) $31 \overline{)245}$

Name: Date: /   /

1 ## Divide:

(1)   41 ⟌ 8 5   (5)   41 ⟌ 1 2 1   (9)   41 ⟌ 2 0 5

(2)   41 ⟌ 8 1   (6)   41 ⟌ 1 6 5   (10)   41 ⟌ 2 3 5

(3)   41 ⟌ 1 2 5   (7)   41 ⟌ 1 6 3

(4)   41 ⟌ 1 2 3   (8)   41 ⟌ 2 0 8

Kamali Academy Division

2 **Divide:**

(1) 41 ⟌ 2 4 8

(2) 41 ⟌ 2 4 6

(3) 41 ⟌ 2 4 4

(4) 41 ⟌ 2 8 0

(5) 41 ⟌ 2 9 0

(6) 41 ⟌ 3 2 0

(7) 41 ⟌ 3 3 0

(8) 41 ⟌ 3 6 0

(9) 41 ⟌ 3 7 0

(10) 41 ⟌ 4 0 0

Name: Date: /   /

1   **Divide:**

(1)   22 | 6 6

(2)   22 | 7 5

(3)   22 | 8 8

(4)   22 | 1 0 0

(5)   22 | 1 1 0

(6)   22 | 1 2 5

(7)   22 | 1 3 2

(8)   22 | 1 4 5

(9)   22 | 1 5 4

(10)   22 | 1 6 5

Kamali Academy Division

2 **Divide:**

(1) 23 ⟌ 46

(2) 23 ⟌ 70

(3) 23 ⟌ 95

(4) 23 ⟌ 120

(5) 23 ⟌ 145

(6) 32 ⟌ 320

(7) 32 ⟌ 70

(8) 32 ⟌ 110

(9) 32 ⟌ 160

(10) 32 ⟌ 200

Name: Date: /   /

1   Divide:

(1) $33\overline{)99}$

(2) $33\overline{)135}$

(3) $33\overline{)165}$

(4) $33\overline{)215}$

(5) $33\overline{)255}$

(6) $42\overline{)120}$

(7) $42\overline{)150}$

(8) $42\overline{)180}$

(9) $42\overline{)210}$

(10) $42\overline{)300}$

2 **Divide:**

(1) 43 ⟌ 1 0 0

(2) 43 ⟌ 1 5 0

(3) 43 ⟌ 2 1 5

(4) 43 ⟌ 2 8 0

(5) 43 ⟌ 3 4 0

(6) 24 ⟌ 8 0

(7) 24 ⟌ 9 6

(8) 24 ⟌ 1 1 0

(9) 24 ⟌ 1 2 0

(10) 24 ⟌ 1 5 0

Name: ___________ Date: ___ / ___ / ___

1 **Divide:**

(1) 34 ⟌ 9 0

(2) 34 ⟌ 1 3 0

(3) 34 ⟌ 1 7 0

(4) 34 ⟌ 2 1 0

(5) 34 ⟌ 2 5 5

(6) 25 ⟌ 1 1 0

(7) 25 ⟌ 1 2 5

(8) 25 ⟌ 1 7 5

(9) 25 ⟌ 2 1 0

(10) 25 ⟌ 2 2 5

 Divide:

(1) 27 | 1 1 0

(2) 27 | 1 3 5

(3) 27 | 1 6 2

(4) 27 | 1 9 0

(5) 27 | 2 1 6

(6) 36 | 1 0 0

(7) 36 | 1 4 4

(8) 36 | 1 6 0

(9) 36 | 1 8 0

(10) 36 | 2 1 5

Name: Date: /   /

1 Divide:

(1) $37 \overline{)115}$

(2) $37 \overline{)150}$

(3) $37 \overline{)185}$

(4) $37 \overline{)230}$

(5) $37 \overline{)259}$

(6) $44 \overline{)132}$

(7) $44 \overline{)180}$

(8) $44 \overline{)220}$

(9) $44 \overline{)280}$

(10) $44 \overline{)352}$

Kamali Academy Division

② Divide:

(1) 45 | 1 8 0

(2) 45 | 2 3 0

(3) 45 | 2 7 0

(4) 45 | 3 1 0

(5) 45 | 3 6 0

(6) 47 | 1 5 0

(7) 47 | 2 0 0

(8) 47 | 2 3 5

(9) 47 | 2 8 2

(10) 47 | 3 3 0

Name: ______ Date: / /

1 Divide:

(1) 54 ⟌ 1 1 0

(2) 54 ⟌ 1 5 0

(3) 54 ⟌ 2 2 0

(4) 55 ⟌ 3 5 0

(5) 55 ⟌ 4 4 0

(6) 63 ⟌ 1 3 0

(7) 63 ⟌ 2 0 0

(8) 63 ⟌ 2 6 0

(9) 64 ⟌ 3 6 0

(10) 64 ⟌ 4 8 0

 Divide:

(1) 56 ⟌ 1 1 0

(5) 57 ⟌ 4 5 0

(9) 66 ⟌ 3 0 0

(2) 56 ⟌ 1 9 0

(6) 65 ⟌ 1 4 0

(10) 66 ⟌ 3 6 0

(3) 56 ⟌ 2 8 0

(7) 65 ⟌ 2 0 0

(4) 57 ⟌ 3 5 0

(8) 65 ⟌ 2 6 0

34 3-Digits ÷ 2-Digits

Name: Date: /   /

1 **Divide:**

(1)   61 ⟌ 1 8 6   (5)   64 ⟌ 4 4 1   (9)   74 ⟌ 5 1 1

(2)   62 ⟌ 1 8 6   (6)   71 ⟌ 2 1 6   (10)   74 ⟌ 6 2 0

(3)   63 ⟌ 3 0 5   (7)   72 ⟌ 2 1 6

(4)   63 ⟌ 4 4 1   (8)   73 ⟌ 3 5 5

      Kamali Academy Division

 Divide:

(1) $81 \overline{)252}$

(2) $82 \overline{)252}$

(3) $82 \overline{)415}$

(4) $83 \overline{)415}$

(5) $83 \overline{)756}$

(6) $84 \overline{)756}$

(7) $91 \overline{)276}$

(8) $92 \overline{)465}$

(9) $93 \overline{)465}$

(10) $93 \overline{)644}$

Name: ___________   Date: ___ / ___ / ___

Divide:

(1) 75 ⟌ 1 5 2

(2) 75 ⟌ 2 2 5

(3) 75 ⟌ 3 7 5

(4) 75 ⟌ 4 5 6

(5) 75 ⟌ 5 6 0

(6) 76 ⟌ 3 0 4

(7) 76 ⟌ 4 5 6

(8) 76 ⟌ 5 0 0

(9) 76 ⟌ 5 2 5

(10) 76 ⟌ 6 0 8

   Kamali Academy Division

 Divide:

(1) 85 | 2 5 5

(2) 85 | 3 4 4

(3) 85 | 4 2 5

(4) 85 | 5 1 6

(5) 86 | 4 2 5

(6) 95 | 3 8 4

(7) 95 | 4 7 5

(8) 95 | 7 0 0

(9) 96 | 4 7 5

(10) 96 | 7 0 0

Name:

Date: / /

1 Divide:

(1) 21 ⟌ 1 3 4 5

(4) 21 ⟌ 1 0 9 7

(7) 31 ⟌ 1 2 3 5

(2) 21 ⟌ 1 4 5 6

(5) 21 ⟌ 1 3 0 0

(8) 31 ⟌ 1 6 7 8

(3) 21 ⟌ 1 5 6 7

(6) 21 ⟌ 1 9 9 0

(9) 31 ⟌ 2 6 0 0

Kamali Academy Division

(1) 41 ⟌ 1 3 4 5   (5) 31 ⟌ 1 2 4 5   (9) 25 ⟌ 1 0 0 4

(2) 41 ⟌ 2 1 2 3   (6) 31 ⟌ 1 2 5 5   (10) 28 ⟌ 1 1 2 5

(3) 41 ⟌ 2 9 8 7   (7) 31 ⟌ 1 5 7 5   (11) 35 ⟌ 1 7 5 4

(4) 41 ⟌ 4 0 0 0   (8) 31 ⟌ 1 8 8 0

4-Digits ÷ 2-Digits

Name: _______________   Date: /   /

① Divide:

(1) 35 | 1 4 3 5

(2) 37 | 1 4 3 5

(3) 39 | 1 4 3 5

(4) 41 | 1 4 3 5

(5) 42 | 2 3 4 5

(6) 44 | 2 3 4 5

(7) 48 | 2 3 4 5

(8) 50 | 2 3 4 5

(9) 51 | 3 4 5 6

Kamali Academy Division

2 **Divide:**

(1) 55 | 4 0 0 0

(2) 56 | 4 0 0 0

(3) 58 | 4 0 0 0

(4) 60 | 4 0 0 0

(5) 39 | 2 3 5 7

(6) 41 | 2 3 5 7

(7) 43 | 3 9 8 1

(8) 51 | 3 9 8 1

(9) 54 | 4 8 5 7

(10) 68 | 4 8 5 7

Name: Date: /   /

1 Divide:

(1) 21 ⟌ 4 8 4 6

(2) 21 ⟌ 6 7 8 9

(3) 21 ⟌ 5 6 7 8

(4) 31 ⟌ 8 9 0 2

(5) 31 ⟌ 6 7 9 9

(6) 31 ⟌ 6 4 2 1

(7) 34 ⟌ 4 9 0 7

(8) 34 ⟌ 5 6 1 4

(9) 34 ⟌ 6 3 2 1

 Divide:

(1) 32 | 6 8 9 0

(2) 32 | 4 0 0 0

(3) 41 | 6 7 8 9

(4) 42 | 6 7 8 9

(5) 42 | 8 4 1 2

(6) 43 | 8 4 1 2

(7) 43 | 8 9 0 2

(8) 44 | 8 9 0 2

(9) 46 | 8 8 0 3

(10) 46 | 8 1 0 5

Name: Date: /   /

Read the problem carefully and solve. Show your work under each question.

The students in Baba Camara's grade have after-school activities. Students can choose between coloring, playing mancala, or basketball.

---

1. 160 students decide to play basketball. These students divide evenly into 8 groups. How many students are in each group?

_______________________ students in each group

2. There were 234 mancala marbles in a pile. 6 students divided them evenly among themselves. How many marbles did each have.

_______________________ marbles

3. Only 5 students decide to color. There are 225 crayons. If each student gets the same number of crayons, how many crayons will each student get?

_______________________ crayons for each

Problem Solving: Dividing by 1-Digit

1. There are 234 summer jobs for lifeguards at the city pools. There will be 3 lifeguards at each city pool. How many city pools are there?

There are _____________ city pools.

2. The Afrikan Girls in Training baked vegan cakes for a community service event. There were 125 different types of cakes. Each baker baked the same number of cakes. If there were 5 bakers, how many cakes did each baker make?

Each baker made _______________ vegan cakes.

3. The Reggae Shop is open 154 hours a week. The shop is open 7 days a week and the same number of hours each day. How many hours each day is the shop open?

The shop is open _________________ hours a day.

Problem Solving: Dividing by 2-Digits

1. A RBG summer camp wants to buy 784 bricks to build a clubhouse. If there are 14 bricks in each box, how many boxes should the camp buy?

____________ boxes

2. Abena's Printing needs to ship 882 programs to Little Afrika. She can fit 21 programs in each box. How many boxes will she need to use?

____________ boxes

3. A new Afrikan clothing website has $4,935 to buy online ads. If each ad costs $15, how many ads can the website purchase?

____________ ads

Problem Solving: Dividing by 2-Digits

1. Sanaa has 2,079 beads to put on the bracelets she is making. It takes 11 beads to make each bracelet. How many bracelets can Sanaa make?

____________ bracelets

2. The Warrior Gun Club budgeted $6,645 to buy new paintball guns for combat training. Each gun costs $15. How many guns will the gun club be able to buy?

____________ guns

3. Kamali International needs to transport 2,668 fans from the parking lot to the football game. If each bus holds 29 people, how many buses should Kamali plan to use?

____________ buses

Problem Solving: Dividing by 2-Digits

1. BlackStar farms produced 8,416 pounds of vegetables. The company wants to put the vegetables into boxes that hold 16 pounds. How many boxes will the vegetables fill?

______________ boxes

2. A ferry needs to transport 330 people across the river to a museum. The ferry can take 22 people on each trip. How many people will the ferry take on its last trip?

______________ people

3. Armah made 837 toy crocodiles and put them in silver tins to share with his family and friends. He put 93 crocodiles in each tin. How many tins was Armah able to fill?

______________ tins

Problem Solving: Dividing by 2-Digits

1. 135 people need to ride the elevator to the top of the monument in Senegal. The elevator can hold 10 people at a time. How many people will be in the elevator on the last trip to the top?

____________ people

2. Mama Oya has 657 gold stickers and wants to give an equal number to each of her 53 students. If Mama Oya has 657 gold stickers and she gives the same number of stickers to each student in the class, how many stickers will be left over?

____________ stickers

3. Garvey Crafts Store needs to ship 97 Afrikan masks across the country. They can fit 36 masks in each truck. How many masks will be in the partially full truck?

____________ masks

Notes:

Notes:

Notes: